SCIENCE 2

Activities

For use with SCIENCE 2
Fifth Edition

bju press
Greenville, South Carolina

Note: The fact that materials produced by other publishers may be referred to in this volume does not constitute an endorsement of the content or theological position of materials produced by such publishers. Any references and ancillary materials are listed as an aid to the student or the teacher and in an attempt to maintain the accepted academic standards of the publishing industry.

SCIENCE 2 Activities
For use with SCIENCE 2 Fifth Edition

Coordinating Writer
Leah R. Thomas, MEd

Writers
Serena Minnick Collins
Debra Harrold White

Consultant
L. Michelle Rosier

Biblical Worldview
Bryan Smith, PhD
Tyler Trometer, MDiv

Academic Oversight
Jeff Heath, EdD
Rachel Santopietro, MEd

Project Editor
Joanna Lynch, MA

Cover & Concept Designer
Emily Heinz

Page Layout
Jennifer Stuhl

Cover Illustrators
Craig Oesterling
Dana Thompson

Art & Design Facilitator
Jim Frasier

Illustrators
Emily Heinz
Kathy Pflug
Carrie Salazar
Dana Thompson
Courtney Godbey Wise

Project Coordinators
Christopher Daniels
Jennifer Ferguson

Permissions
Ashleigh Schieber
Kathleen Thompson
Elizabeth Walker

Photo credits appear on page 253–54.

ISBN 978-1-62856-609-3

15 14 13 12 11 10 9 8 7 6 5

Looking Ahead

Name ______________________________

Circle the answer.

1. The study of God's world is called ___. science senses
2. A person who uses his senses to observe God's world is called a ___. pathfinder scientist
3. The way a person thinks about the world is called a ___. feeling worldview
4. The engineering ___ ___ is the steps that an engineer uses to solve real-life problems. design process master plan

Match the science inquiry skill with the picture.

_____ 5. classify

_____ 6. measure

_____ 7. infer

_____ 8. predict

A

B

C

D

Match the description with the science tool.

_____ 9. measures how long something is

_____ 10. measures how much space something takes up

_____ 11. measures how much matter is in something

A 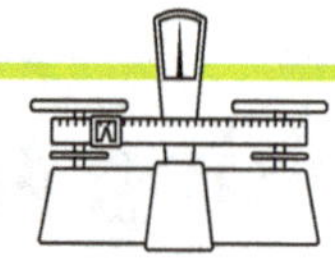

B 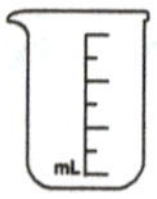

C 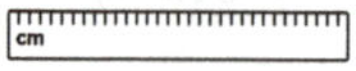

Decide which step of the science investigation is being done. Mark the answer.

12. Jamie wants to know if a cup of ice in a lunch bag will melt faster or slower than a cup outside a lunch bag.

   ○ Problem ○ Procedure

13. Marcy follows the steps in the investigation.

   ○ Procedure ○ Hypothesis

14. Alex looks at the ice and lunch box. He writes a statement to be tested.

   ○ Problem ○ Hypothesis

15. Lydia decides the results of her test agree with her hypothesis.

   ○ Observations ○ Conclusions

Write About It

Two scientists are looking at the same dinosaur bones. Mr. Henning explained that man and the dinosaurs were created on the same day. Miss Frank said the dinosaurs lived many, many years before man.

Explain why the scientists look at the dinosaur bones differently.

____

____

____

Worldview

Name ______________________

Complete the web about worldview.

God created a ______________ world.

Sin brought death and ______________.

God gave people the job of ______________ for the earth.

Biblical

A person believes the Bible is true and allows the ______________ to guide his decisions.

Worldview

is the way a person thinks about the ______________.

Evolution

A person believes the earth and space formed by chance without ______________.

Everything happened ______________ over a long time.

A tiny speck went ______________!

Life began to ______________ many, many years ago.

Science and Scientists

Complete the definitions. If needed, use the Glossary.

1. **Science** is the study of God's ________________.

2. A **scientist** is a person who uses his ________________ to observe God's world.

3. A **worldview** is the way a person ________________ about the world.

Write *T* if the statement is true. If the statement is false, write the correction for the underlined words.

________________ 4. You can use your <u>mind</u> to solve problems and study God's world.

________________ 5. When a scientist helps other people, he is obeying <u>man's</u> command to love our neighbors.

________________ 6. All scientists have a <u>worldview.</u>

Write *C* if the statement is a belief of Creation. Write *E* if the statement is a belief of evolution.

______ 7. God made the world perfect. The earth changed after Adam sinned.

______ 8. The earth and space formed by chance.

______ 9. Because of sin, pain and death entered into the world.

______ 10. Everything happened slowly over a long, long time.

Answer the question. Use the Bible's teaching from Genesis 1:28 in your answer.

Genesis 1:28

God gave people the job of caring for the earth.

11. What is the purpose for what a scientist does?

____

____

____

Inquiry Skills

Name ______________________

Complete the web about the inquiry skills.

Science Inquiry Skills

skills or actions to help a scientist gather and use

____

1 Observe

to use your

____

to find out something

2 Classify

to group things that are

____

3 Measure

to find the size,

______________________, or amount of something

4 Predict

to make a careful

____

at what may happen after observing

5 Infer

to use what you know to tell

____

things happen

6 Communicate

to ______________________

or ______________________

what you know

Five Fingers

Complete the graphic organizer about science tools.

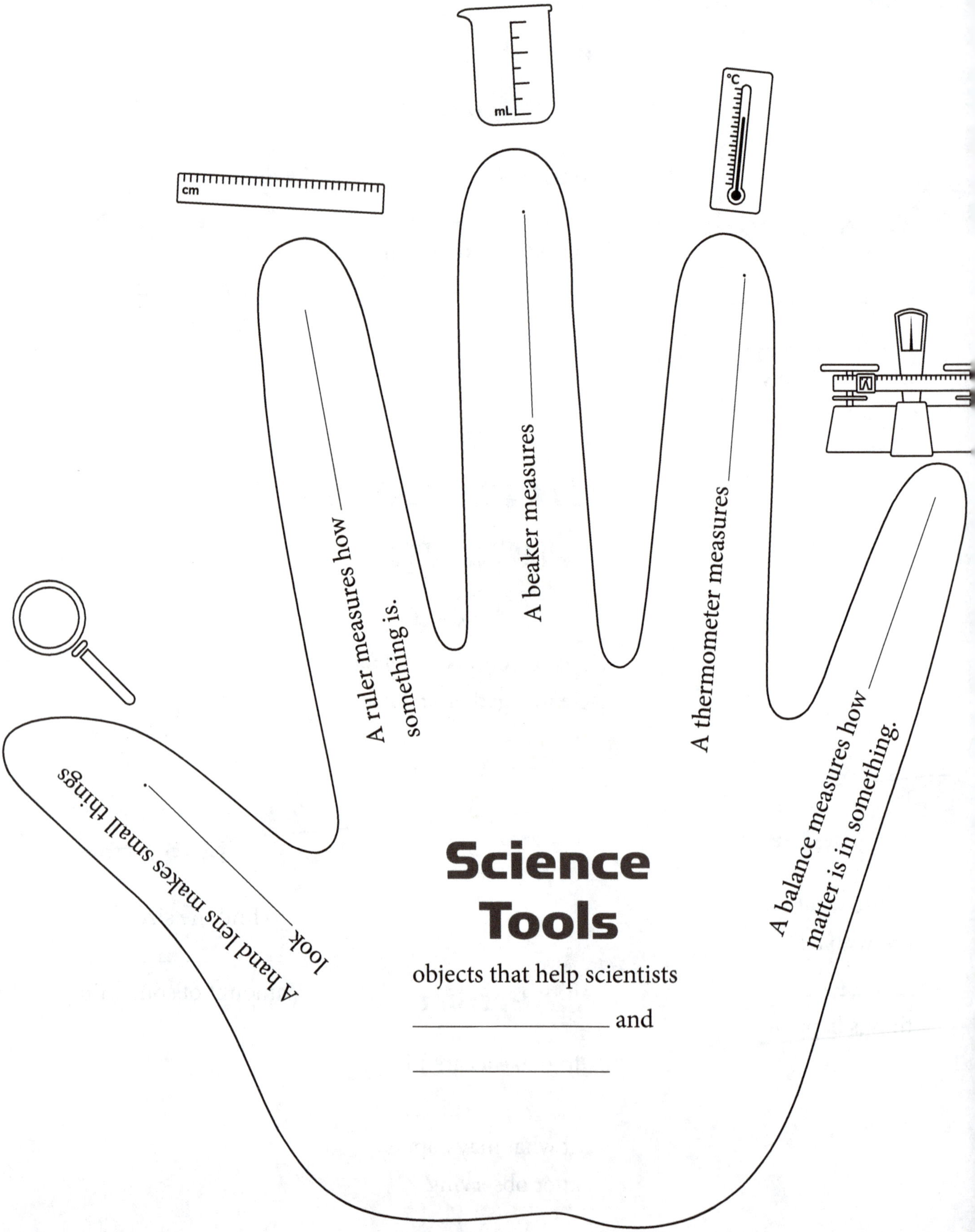

Skills and Tools

Name ______________________________

Complete the section.

1. A **science inquiry skill** is a skill or action to help a scientist gather and use ________________.

2. You can use science tools to help you ________________ and ________________.

3. Why is it important to measure correctly? Use the Bible's teaching from Genesis 1:28 in your answer. ____
____
____
____

Genesis 1:28

God gave people the job of caring for the earth.

Match each science inquiry skill with its picture.

A

B
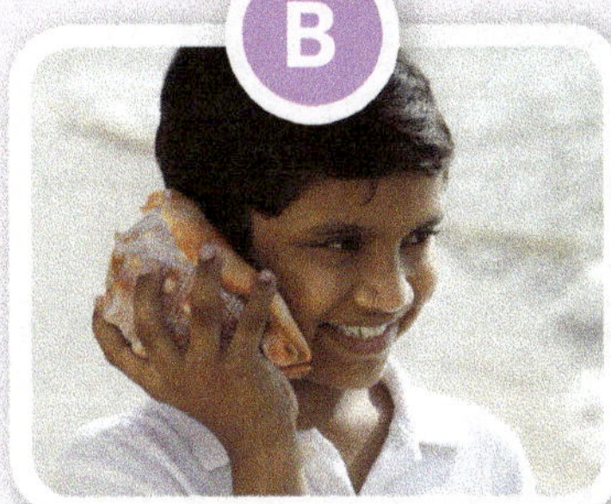

C

_____ 4. observe

_____ 5. classify

_____ 6. measure

D

E

F

_____ 7. predict

_____ 8. infer

_____ 9. communicate

Match each tool with what it is used for.

______ 10.

______ 11.

______ 12.

______ 13.

______ 14.

A measures volume

B measures temperature

C measures how much matter is in something

D makes small things look bigger

E measures how long something is

INVESTIGATION

Keeping Cool

Name ______________________________

Inquiry Skills
- Measure
- Infer

An **investigation** is a scientific test to solve a problem.

Each investigation will direct attention to two or three **inquiry skills**.

Problem

An investigation begins with a **problem**. The problem is a *question* that needs to be answered.

Does color affect temperature?

Hypothesis

A **hypothesis** is a possible answer to the question.

Complete the hypothesis.

The thermometer in black paper will change temperature more / less than the thermometer in white paper.

Materials

The **materials** are the supplies needed to do the investigation.

Materials
- ☐ ½ sheet of black construction paper
- ☐ ½ sheet of white construction paper
- ☐ 2 thermometers

Procedure

The **procedure** is the steps of an investigation.

Follow the steps to do the investigation.

- [ ] 1. Fold the black paper in half. Tape the long edge and one short edge to make a sleeve as shown.

- [ ] 2. Fold and tape the white paper the same way.
- [ ] 3. Read the temperatures on both thermometers.
- [ ] 4. Write them as the beginning temperatures on the Observations chart.
- [ ] 5. Place one thermometer in each paper sleeve.
- [ ] 6. Place the thermometers beside each other in a sunny place outside.
- [ ] 7. Wait about an hour.
- [ ] 8. Read the temperatures.
- [ ] 9. Write the ending temperatures for both thermometers in the chart.
- [ ] 10. Find the difference between the beginning temperature and the ending temperature for both thermometers.

Science Inquiry Skills

Name ______________________________

Observe | to use your senses to find out something

Classify | to group things that are alike

Measure | to find the size, number, or amount of something

Infer | to use what you know to tell why things happen

Predict | to make a careful guess at what may happen after observing; using what you know to say what may happen

Communicate | to share what you know; to give information about something

Science Safety Tips

1. Listen to your teacher's directions.
2. Use tools the way they were made to be used.
3. Handle supplies with care.
4. Wear safety goggles when needed.
5. Tell your teacher about accidents.
6. Wait your turn.

Looking Ahead

Name ____________________

Write an *X* in the first column if you have never heard the word before.
Write an *X* in the second column if you have heard the word before.
In the last column, write what you think each word means.

Words to Know

Word	Have Not Heard Before	Have Heard Before	What I Think It Means
matter			
mass			
property			

Answer the question.

What are some properties of matter? ____________________

____

____

____

____

Chapter 2

Bible Verses

Genesis 1:1

In the beginning God created the heaven and the earth.

Genesis 1:11–13

11 And God said, Let the earth bring forth grass,
the herb yielding seed, and the fruit tree yielding
fruit after his kind, whose seed is in itself, upon
the earth: and it was so.
12 And the earth brought forth grass, and herb
yielding seed after his kind, and the tree yielding
fruit, whose seed was in itself, after his kind: and
God saw that it was good.
13 And the evening and the morning were the
third day.

INVESTIGATION

How Slow Is the Flow?

Name ______________________________

Inquiry Skills

- Predict
- Observe

Materials

- ▢ baking sheet
- ▢ masking tape
- ▢ marker
- ▢ 3 books
- ▢ 2 prepared cups
- ▢ hand towel

Problem

Which liquid flows the slowest?

Hypothesis

The liquid that flows the slowest is ________________.

Procedure

- ▢ 1. Use masking tape to mark a start line and a finish line on the baking sheet. Label one line *Start* and the other line *Finish*.
- ▢ 2. Raise the end of the baking sheet marked *Start* by placing the books under that end.
- ▢ 3. Under Observations, identify the liquids used.
- ▢ 4. Ask a friend to help you. Tip both cups of liquid at the same time at the *Start* line.
- ▢ 5. Observe which liquid finishes last.
- ▢ 6. Circle the liquid that flowed the slowest.

Observations

First liquid ______________________

Second liquid ______________________

Conclusions

1. Did the liquids flow at the same speed? ○ yes ○ no
2. Which liquid is thicker? ______________________
3. Explain how you know which liquid is thicker. ______________________

Looking Ahead

Name ______________________________

Write an *X* in the first column if you have never heard the word before.
Write an *X* in the second column if you have heard the word before.
In the last column, write what you think each word means.

Words to Know

Word	Have Not Heard Before	Have Heard Before	What I Think It Means
evaporation			
condensation			
mixture			

Answer the question.

How does a solid change when it is heated? ______________________________

____

____

____

Chapter 3

Bible Verses

Genesis 1:1

In the beginning God created the heaven and the earth.

Psalm 147:7–8

7 Sing unto the LORD with thanksgiving;
sing praise upon the harp unto our God:
8 Who covereth the heaven with clouds,
who prepareth rain for the earth, who
maketh grass to grow upon the mountains.

Matthew 22:39b

Thou shalt love thy neighbour as thyself.

Changing Matter

Name ______________________________

Write the answer.

1. As stated in Genesis 1, where did all matter come from? ______________________________

____

2. What are the three states of matter?

____

____

____

Mark the answer.

3. What tool is used to measure the temperature of matter?

○ balance ○ thermometer

Circle the answer.

4. Heat is added to water. The temperature of the water goes up / down.

5. Heat is removed from an apple pie. The temperature of the pie goes up / down.

Write the letters of two correct answers.

_____ 6. When a solid melts, it ____.

A changes shape

B keeps its shape

C becomes a liquid

Look at the pictures. Write the answer.

_____ 7. Carl and Andy built a snowman. The sun came out. The temperature went up. What happened to the snowman?

A The matter changed from a liquid to a solid.

B The matter changed from a solid to a liquid.

C The matter did not change.

INVESTIGATION

Reversible or Irreversible Changes?

Name ____________________

Inquiry Skills

- Predict
- Observe
- Infer

Materials

- ☐ 4–5 broken crayons, all of one color
- ☐ clear glass heat-resistant jar
- ☐ hot plate
- ☐ saucepan
- ☐ water
- ☐ tongs
- ☐ hot pad
- ☐ silicone mold
- ☐ raw egg
- ☐ bowl
- ☐ slotted spoon

reversible change: a change that can be undone

irreversible change: a change that cannot be undone

Problem

Are the changes caused by heating and cooling crayons reversible or irreversible?

Are the changes caused by heating and cooling an egg reversible or irreversible?

Hypothesis

The changes caused by heating crayons are reversible / irreversible.

The changes caused by heating an egg are reversible / irreversible.

Procedure

Fill in the Observations chart when the procedure states Observe and identify.

CAUTION

This Investigation is to be done with adult supervision. The students will follow each step of the procedure as an adult does the Investigation.

Use caution around the hot plate and its cord.

- ☐ 1. Place several broken crayon pieces into the clear glass jar.
- ☐ 2. **Observe and identify** the *state* of a crayon. (Circle one state.)
- ☐ 3. **Observe and identify** the *properties* of a crayon.
- ☐ 4. Put water in the saucepan and set it on the hot plate.
- ☐ 5. Put the jar of crayons upright in the water. Heat the crayons until they melt.
- ☐ 6. Use tongs to steady the jar, as needed, while the crayons are melting and to lift the jar of melted crayons out of the boiling water.
- ☐ 7. Pour the melted crayons into the silicone mold.
- ☐ 8. **Observe and identify** the *state* of the crayons. (Circle one state.)
- ☐ 9. **Observe and identify** the *properties* of the crayons in the mold.
- ☐ 10. Allow the crayons in the mold to cool.
- ☐ 11. While waiting for the crayons to cool, crack an egg into a bowl. Break the yolk.
- ☐ 12. **Observe and identify** the *state* of the egg. (Circle one state.)
- ☐ 13. **Observe and identify** the *properties* of the egg.
- ☐ 14. Keep a gentle boil in the saucepan of water.
- ☐ 15. Pour the egg into the water. Allow it to remain in the boiling water for a few minutes.
- ☐ 16. Remove the egg from the water with a slotted spoon.

INVESTIGATION

Reversible or Irreversible Changes?

Name ______________________________

- ☐ 17. **Observe and identify** the *state* of the egg after heat has been added. (Circle one state.)
- ☐ 18. **Observe and identify** the *properties* of the egg after heat has been added.
- ☐ 19. Allow the egg to cool.
- ☐ 20. **Observe and identify** the *state* of the cooled, cooked egg. (Circle one state.)
- ☐ 21. **Observe and identify** the *properties* of the cooled, cooked egg.
- ☐ 22. After the crayon has cooled, remove it from the mold.
- ☐ 23. **Observe and identify** the *state* of the cooled crayon. (Circle one state.)
- ☐ 24. **Observe and identify** the *properties* of the cooled crayon.

Observations

Crayons

Starting State	Starting Properties	Heat Added		Heat Removed	
		State	Properties	State	Properties
solid liquid gas		solid liquid gas		solid liquid gas	

Egg

Starting State	Starting Properties	Heat Added		Heat Removed	
		State	Properties	State	Properties
solid liquid gas		solid liquid gas		solid liquid gas	

INVESTIGATION

Reversible or Irreversible Changes?

Name ___________________________

Conclusions

Read the definitions of *reversible change* and *irreversible change* on page 51. Use what you know from your observations to draw your conclusions.

1. How can you change the state of a crayon from a solid to a liquid?

   ○ add heat ○ remove heat

2. Does a crayon that has been heated and cooled have the same properties as the original crayon?

   ○ yes ○ no

3. Is the change to the crayons reversible or irreversible?

   ○ reversible ○ irreversible

4. How do you know that the change to the heated crayons is reversible or irreversible?

   ____

   ____

   ____

5. How can you change the state of an uncooked egg from a liquid to a solid?

   ○ add heat ○ remove heat

6. Does an egg that has been heated and cooled have the same properties as the original egg?

   ○ yes ○ no

7. When an egg is heated, is the change reversible or irreversible?

   ○ reversible ○ irreversible

8. How do you know that the change to the heated egg is reversible or irreversible?

   ____

   ____

Looking Ahead

Name ______________________________

Write an *X* in the first column if you have never heard the word before.
Write an *X* in the second column if you have heard the word before.
In the last column, write what you think each word means.

Words to Know

Word	Have Not Heard Before	Have Heard Before	What I Think It Means
force			
gravity			
motion			

Write the answer.

What kind of things are attracted to a magnet? ______________________________

____

Chapter 4

Bible Verses

Genesis 1:14

And God said, Let there be lights in the firmament of the heaven to divide the day from the night; and let them be for signs, and for seasons, and for days, and years:

Genesis 1:16

And God made two great lights; the greater light to rule the day, and the lesser light to rule the night: he made the stars also.

Genesis 1:27

So God created man in his own image, in the image of God created he him; male and female created he them.

Genesis 1:28

And God blessed them, and God said unto them, Be fruitful, and multiply, and replenish the earth, and subdue it: and have dominion over the fish of the sea, and over the fowl of the air, and over every living thing that moveth upon the earth.

Forces Organizer

Name ____________________

Complete the web about forces.

Example ____________________

____

____

____

1 ____________________

a force that ____________________ or ____________________ motion

Example ____________________

____

____

____

2 ____________________

the force that ____________________ all things toward the ____________________ of the earth

Kinds of Forces

3 ____________________ ____________________

the force of a ____________________

Example ____________________

____

____

Friction

Name ______________________________

Circle the picture that answers the question.
Then complete the reason why you circled the picture.

1. Which pair of boots would cause the most friction when you walk?

I circled the picture because it shows boots with rough / smooth soles.

2. Which surface does not have very much friction?

I circled the picture because it shows a rough / smooth surface.

3. Which surface has more friction?

I circled the picture because it shows a rough / smooth surface.

4. Which picture shows someone adding friction to a surface?

I circled the picture because it shows adding a rough / smooth surface to make more friction.

EXPLORATION

All Fall Down

Name ______________________________

Inquiry Skills

- Observe
- Infer

Materials

- ▢ thick towel
- ▢ Ping-Pong® ball
- ▢ softball
- ▢ golf ball
- ▢ piece of paper

Problem

Do some objects fall faster than others?

Procedure

▢ 1. Put the thick towel on the floor.

▢ 2. Drop the Ping-Pong® ball and softball from the same height at the same time above the towel.

▢ 3. Circle what happened.

The Ping-Pong ball hit first.

The softball hit first.

Both balls hit at the same time.

▢ 4. Drop the golf ball and the softball from the same height at the same time above the towel.

▢ 5. Circle what happened.

The golf ball hit first.

The softball hit first.

Both balls hit at the same time.

▢ 6. Crumple the piece of paper into a ball about the same size as the Ping-Pong ball.

- [ ] 7. Drop the paper ball and the Ping-Pong ball from the same height at the same time above the towel.

- [ ] 8. Circle what happened.

  The paper ball hit first.

  The Ping-Pong ball hit first.

  Both balls hit at the same time.

Conclusions

1. Were all the balls the same size?	O yes	O no
2. Were all the balls the same weight?	O yes	O no
3. Were all the balls about the same shape?	O yes	O no
4. When the balls were dropped from the same height at the same time, did they fall at the same speed?	O yes	O no
5. Did any ball fall faster than another?	O yes	O no

6. What can you infer about any two objects that are dropped from the same height at the same time? ______________________________

____

INVESTIGATION

Magnetic Attraction

Name ______________________________

Inquiry Skills
- Predict
- Classify

Materials
- ▢ plastic button
- ▢ eraser
- ▢ nail
- ▢ metal paper clip
- ▢ penny
- ▢ staples
- ▢ 3 other objects
- ▢ small paper bag
- ▢ magnet

Problem

Which items are attracted to a magnet?

Hypothesis

You will complete your hypothesis later in the Investigation.

Procedure

- ▢ 1. Choose 3 objects to be tested. Write the names of the 3 objects on the chart under Observations.
- ▢ 2. Examine each object. Record your prediction on the chart. Circle *yes* if you expect the object to stick to the magnet. Circle *no* if you do not expect the object to stick to the magnet.
- ▢ 3. Place all the objects in a paper bag.
- ▢ 4. Place the magnet inside the bag and gently shake the bag.
- ▢ 5. Pull out the magnet. Observe the objects that stuck to it.
- ▢ 6. Record your results. Circle *yes* for the objects that stuck to the magnet.
- ▢ 7. Repeat steps 4–6 until there are no objects left in the bag that stick to the magnet.
- ▢ 8. Put all the objects that stuck to the magnet together.
- ▢ 9. Record your results. Circle *yes* for the objects that stuck to the magnet.
- ▢ 10. Remove any remaining objects from the bag. Keep them in a separate group.
- ▢ 11. On the chart, circle *no* for the objects that did not stick to the magnet.

Observations

Item	Prediction: Will It Attract?	Observation: Did It Attract?	Did Your Results Match Your Prediction?
button	yes no	yes no	yes no
eraser	yes no	yes no	yes no
nail	yes no	yes no	yes no
paper clip	yes no	yes no	yes no
penny	yes no	yes no	yes no
staples	yes no	yes no	yes no
	yes no	yes no	yes no
	yes no	yes no	yes no
	yes no	yes no	yes no

Conclusions

1. Did your results match your predications? Complete the last column in the chart.
2. Examine the objects that stuck to the magnet. What can you summarize about the objects that stuck to the magnet? ______________________________
3. Look at the two groups of objects. How could you classify the objects you put in the bag? ______________________________
4. How can you use what you learned about magnetic objects? ______________________________

____

____

Study Guide

Name ______________________________

Mark the answer.

13. Look at these pictures. Which magnets will attract each other?

○ 

○ 

○ 

Write the answer.

14. Why do you learn about force? Include one example with your answer. ______________________________

____

____

You push off the ground at the top of a hill. You ride your scooter all the way down. Gravity causes the scooter to roll down the hill. At the bottom of the hill, your scooter rolls to a stop.

Explain what force causes the scooter to stop at the bottom of the hill.

15. ______________________________

____

____

____

Looking Ahead

Name ______________________________

Write an *X* in the first column if you have never heard the word before.
Write an *X* in the second column if you have heard the word before.
In the last column, write what you think each word means.

Words to Know

Word	Have Not Heard Before	Have Heard Before	What I Think It Means
solar system			
axis			
orbit			

Answer the question.

Why does the earth have different seasons? ______________________________

____

Chapter 5

Bible Verses

Genesis 1:14, 16–18

14 **And God said, Let there be lights in the
firmament of the heaven to divide the day from
the night; and let them be for signs, and for
seasons, and for days, and years:**

16 **And God made two great lights; the greater
light to rule the day, and the lesser light to rule
the night: he made the stars also.**

17 **And God set them in the firmament of the
heaven to give light upon the earth,**

18 **And to rule over the day and over the night,
and to divide the light from the darkness: and
God saw that it was good.**

Genesis 8:22

While the earth remaineth, seedtime and harvest, and cold and heat, and summer and winter, and day and night shall not cease.

What's the Difference?

Name ______________________

Complete the chart to compare and contrast evolution and Creation.

Evolution	Creation
1. The solar system began with a __________ __________.	2. The solar system was created by __________.
3. The solar system took many, many __________ to form.	4. God made the solar system in the __________ days of Creation.
5. The order in the solar system happened by __________.	6. The order in the solar system shows God's __________.

Answer the questions.

7. Does the Bible agree with Creation or evolution? __________

8. How do you know that a belief in Creation agrees with the biblical worldview? __________

____

____

____

Looking Ahead

Name ______________________

Write an *X* in the first column if you have never heard the word before.
Write an *X* in the second column if you have heard the word before.
In the last column, write what you think each word means.

Words to Know

Word	Have Not Heard Before	Have Heard Before	What I Think It Means
landform			
volcano			
earthquake			

Answer the question.

What are the three layers of the earth?

____

____

____

Chapter 6

Bible Verses

Genesis 1:9–10

9 And God said, Let the waters under the heaven be
gathered together unto one place, and let the dry
land appear: and it was so.
10 And God called the dry land Earth; and the
gathering together of the waters called he Seas: and
God saw that it was good.

Genesis 1:27

So God created man in his own image, in the image of God created he him; male and female created he them.

Matthew 22:37–39

37 Jesus said unto him, Thou shalt love the Lord thy
God with all thy heart, and with all thy soul, and with
all thy mind.
38 This is the first and great commandment.
39 And the second is like unto it, Thou shalt love thy
neighbour as thyself.

Study Guide

Name ___________________________

Write the answer.

21. According to Genesis 1, where did dry land and water on the earth come from?

____

22. A scientist who studies the nonliving parts of the earth is a _______________.

23. How can scientists infer, or draw a conclusion about, what layers are inside the earth?

____

____

____

____

____

24. Why can scientists only infer what layers are inside the earth? _______________________________

____

____

____

EXPLORATION

The Earth's Layers

Name ______________________

Inquiry Skills
- Measure
- Infer

Materials

- ▢ red, yellow, and green modeling clay
- ▢ ruler
- ▢ thread, 30 cm long
- ▢ red, yellow, and green crayons or colored pencils

Purpose

I want to make a model of the layers of the earth.

Procedure

- ▢ 1. Make a ball of red clay that is 3 centimeters (cm) across.
- ▢ 2. Roll out a thick layer of yellow clay.
- ▢ 3. Press the layer of yellow clay around the red ball. Add the yellow clay until the ball is almost 6 cm across.
- ▢ 4. Roll out a thin layer of green clay. Use the green clay to completely cover the yellow clay ball.
- ▢ 5. Hold a length of thread (30 cm long) tightly between both hands. Slice through the middle of the clay ball with the thread.
- ▢ 6. Use the circle diagram to draw and color the layers of your model.
- ▢ 7. Label the layers of the earth on the diagram.

Draw and color the layers of the earth that you observe after you slice the clay model apart. Label the layers of the earth.

Conclusions

1. Which layer of the earth is the thinnest? ____________________

2. Which layer of the earth is in the center? ____________________

3. Why did you have to measure the thickness of the clay as you made your model?

____

____

____

Name ______________________

Write About It

Penny is reading a picture book about weathering and erosion. The book says that weathering and erosion usually happen very slowly. Penny remembers learning something different in science class. But she cannot remember all that she has learned.

Help Penny remember how fast weathering and erosion can happen. Read Genesis 6–8 and Psalm 104:5–9 in the Bible to help you with your answer.

____

____

____

____

____

Looking Ahead

Name ______________________________

Write an *X* in the first column if you have never heard the word before.
Write an *X* in the second column if you have heard the word before.
In the last column, write what you think each word means.

Words to Know

Word	Have Not Heard Before	Have Heard Before	What I Think It Means
natural resources			
conserve			
product			

Write the answer.

What is an object that you can recycle? ______________________________

____

Chapter 7

Bible Verse

Genesis 1:28

And God blessed them, and God said unto them, Be fruitful, and multiply, and replenish the earth, and subdue it: and have dominion over the fish of the sea, and over the fowl of the air, and over every living thing that moveth upon the earth.

Natural Resources

Name ______________________________

Natural Resources

anything in ________ that God made for people to ________

1. conserve ________

2. conserve ________

3. conserve ________

4. conserve ________

5. conserve ________

6. conserve ________

Made by God

Name ______________________________

Complete the crossword puzzle.

AIR	CONSERVE	EROSION	PLANTS
RESOURCES	SOIL	TREES	WATER

Across

3. a natural resource that you drink
4. a natural resource that gives animals and people food
7. used by every living thing to supply its needs
8. a natural resource that helps plants grow

Down

1. the moving of soil from one place to another
2. means to not waste or to use something wisely
5. a natural resource that is used by living things to breathe
6. a natural resource that takes years to grow

Study Guide

Review

Name ______________________

Match each definition with the word. If needed, use the Glossary.

_____ 1. anything in nature that God made for people to use

_____ 2. to use something wisely or to not waste

_____ 3. the moving of rocks, soil, or sand on the earth's surface

_____ 4. a natural resource formed from the remains of plants and animals that were buried quickly

_____ 5. a scientist who protects the wildlife, plants, water, and soil

_____ 6. anything that makes the air, water, or land dirty

A conserve
B erosion
C fossil fuels
D natural resource
E natural resource manager
F pollution

Fill in the blank. One word will be used two times.

erosion	manage	obey	off	wisely

7. One way to conserve air is to ______________ government rules to keep our air clean and healthy.

8. One way to conserve water is to turn ______________ the water when you are finished.

9. One way to conserve soil is to prevent ______________.

10. One way to conserve plants is to use them ______________.

11. One way to conserve animals is to obey God's command to ______________ the earth.

12. One way to conserve fossil fuels is to turn ______________ the light when you leave a room.

Write *T* if the statement is true. If the statement is false, write the correction for the underlined words.

______________ 13. Trees can be conserved by <u>replacing</u> the ones cut down.

______________ 14. We use fossil fuels to produce <u>energy</u>.

______________ 15. One problem <u>soil</u> can cause is pollution.

Write the answer.

16. Name six natural resources.

______________ ______________ ______________

______________ ______________ ______________

17. Name three kinds of fossil fuels.

______________ ______________ ______________

Write the answer. Use the Bible's teaching found in Genesis 1:28.

18. Why should people conserve natural resources? ______________________________

____

19. What are two reasons that people need to use fossil fuels wisely?

____

____

Study Guide

Name ______________________________

Write About It

After each rain Mr. Burns swept dirt from his driveway. He decided to buy low-growing plants. He prepared the soil. Then he put the plants on the hill above the driveway. As the plants grew, they covered the bank. Mr. Burns had less dirt to sweep up.

Explain what is happening to the hill above Mr. Burns's driveway. Then explain why Mr. Burns bought plants and put them on the hill above his driveway.

____

____

____

____

Products

Name ______________________

Complete the definition. If needed, use the Glossary.

1. A **product** is something made by ______________, machines, or a natural process.

Key

A = animal **S** = soil

F = fossil fuels **W** = water

P = plants

Decide which natural resource the product is made from. Use the key to write a letter on the square.

fruit

leather belt

wooden porch swing

washing hands

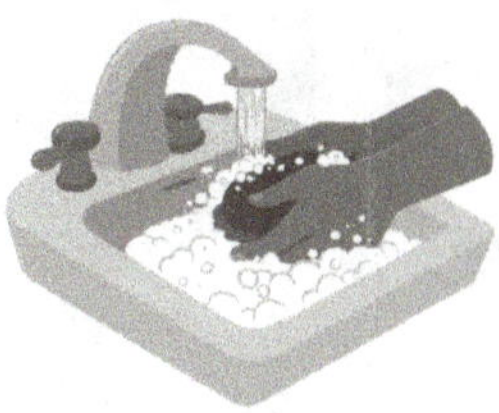

pork chop

concrete statue

wool sweater

feather pillow

drinking water

gasoline

paper towels

brick patio

plastic bag

book

glass window

oil can

Sell It!

Choose a product made from a natural resource. Design a poster in the space below to advertise or sell that product.

Study Guide

Name ______________________________

Review

Complete the definitions.

1. Something made by people, machines, or a natural process is called a ______________.
2. When we use something again we ______________ it.
3. When we use less of something we ______________ it.
4. When we make something new out of an object we ______________ it.

Write the three *R*s that help us remember to conserve our natural resources. Give another example of each one.

5 ______________

Example: reuse an old can for pencils

____

____

____

6 ______________

Example: use cloth bags instead of plastic ones ______________

____

____

7 ______________

Example: recycle cardboard

____

____

Write the natural resource each product is made from. One resource will be used two times.

animals	fossil fuels	plants	soil	water

<table>
<tr>
<td>

milk

8. ________________

</td>
<td>

showering

9. ________________

</td>
<td>

birthday card

10. ________________

</td>
</tr>
<tr>
<td>

brick post

11. ________________

</td>
<td>

wheat bread

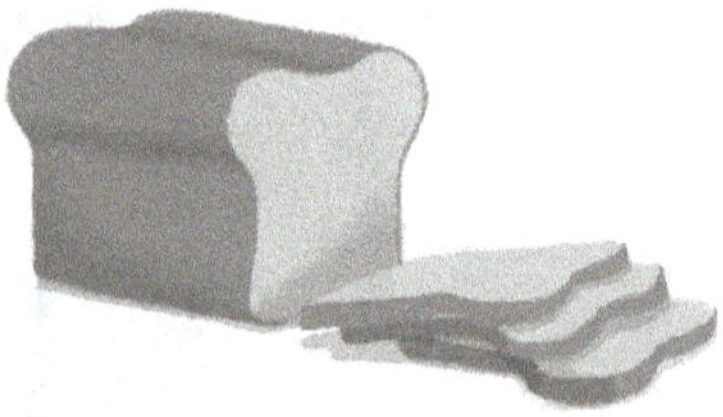

12. ________________

</td>
<td>

electricity

13. ________________

</td>
</tr>
</table>

Write the answer. Use the Bible's teaching from Genesis 1:28.

14. How does using the three *R*s help people obey God? ________________

____

____

____

____

Think About It!

Name ______________________________

Answer the questions.

1. Your little sister has an old plastic wading pool. What are two other ways you could reuse that pool? ______________________

____

____

2. What is one way you can reduce the amount of electricity you use? ________________

____

____

3. What is something you can recycle? ______

____

____

EXPLORATION

Recycled Paper

Name ______________________________

Inquiry Skills

- Infer
- Measure

Materials

- ☐ newspaper
- ☐ water
- ☐ beaker
- ☐ potato masher
- ☐ large plastic dishpan
- ☐ wire screen
- ☐ dishtowel

Purpose

I want to recycle newspapers to make new paper.

Procedure

☐ 1. Tear several sheets of newspaper into small pieces.

☐ 2. Pour 600 milliliters (mL) of water into a dishpan. Add the pieces of newspaper. Make sure each piece gets wet.

☐ 3. Mash and stir the paper with the potato masher for several minutes.

☐ 4. Add 3000 mL of water to the dishpan. Use the potato masher to spread it evenly through the water.

☐ 5. Press the wire screen to the bottom of the dishpan. Gently move the screen back and forth.

☐ 6. Slowly lift the screen out of the dishpan. Hold the screen over the dishpan for about 1 minute to let the extra water drain off.

☐ 7. Set the screen on top of some extra newspaper. Cover with a towel. Press down on the towel to squeeze more water from the screen and pulp.

☐ 8. Remove the towel. Gently peel off your new paper. Set it aside to dry.

Conclusions

1. How is the new paper different from the old newspaper? ______________________

____

2. What could you use this new paper for? ______________________

____

— Matthew 22:37–39 —

God's first commandment is to love God with all my heart, soul, and mind. God's second commandment is to love my neighbor.

3. How does recycling newspaper and other things help you obey God's commandments in Matthew 22:37–39? ______________________

____

____

____

____

____

Looking Ahead

Name ____________________

Write an *X* in the first column if you have never heard the word before.
Write an *X* in the second column if you have heard the word before.
In the last column, write what you think each word means.

Words to Know

Word	Have Not Heard Before	Have Heard Before	What I Think It Means
life cycle			
need			
seedling			

Answer the question.

What are some things plants must have to grow? ____________________

____

____

____

Chapter 8

Bible Verses

Genesis 1:11–13

11 And God said, Let the earth bring forth grass,
the herb yielding seed, and the fruit tree yielding
fruit after his kind, whose seed is in itself, upon
the earth: and it was so.
12 And the earth brought forth grass, and herb
yielding seed after his kind, and the tree yielding
fruit, whose seed was in itself, after his kind: and
God saw that it was good.
13 And the evening and the morning were the
third day.

Genesis 3:17–18a

17 And unto Adam he said, Because thou hast
hearkened unto the voice of thy wife, and hast
eaten of the tree, of which I commanded thee,
saying, Thou shalt not eat of it: cursed is the
ground for thy sake; in sorrow shalt thou eat of it
all the days of thy life;
18a Thorns also and thistles shall it bring forth to
thee.

Study Guide

Name ____________________

Write About It

A boy named Abram planted flowers in his garden. Every other day Abram checked on his flowers. One day he noticed weeds growing near the flowers. The weeds were making it hard for the flowers to grow. He pulled out each weed. It took a long time, but the flowers were no longer crowded by the weeds. He asked his friend Beck why the weeds were growing in his garden.

Help Beck to explain why weeds were growing in the garden and crowding the flowers. Use Genesis 3:17–18 to help you answer Abram's question.

____

____

____

____

____

Study Guide

Name ______________________

Review

Complete the definitions. If needed, use the Glossary.

1. A **life cycle** is all the ______________ of the life of a living thing.
2. To **sprout** is when a plant starts to ______________ from a seed.
3. A **seedling** is a ______________ plant.

Fill in the blanks.

4. Plants were created to make more plants ______________ ______________ ______________.

Mark two answers.

5. To sprout, a seed needs the nutrients in the stored food, and the right amount of ___ and ___.

O soil O space O water O warmth

Complete the diagram of the parts of a seed.

6. __________ __________

7. __________ __________

8. __________ __________

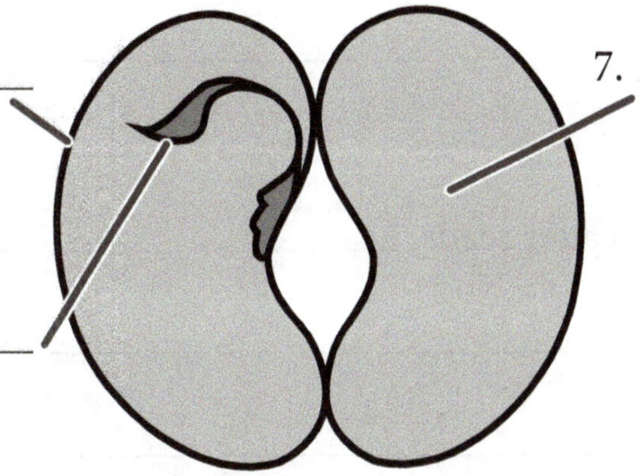

Mark the answer.

9. Why do some plants have seeds?

O to make more of the same kind of plants

O to make more of different kinds of plants

Answer the question.

10. What are the three steps of the life cycle of a plant?

____

____

____

Look at the pictures. Write the way the seeds can travel.

animal **water** **wind**

11. ______________

12. ______________

13. ______________

Write the answer.

14. Explain how animals scatter seeds. ______________________________________

____

____

____

Looking Ahead

Name ___________________________

Write an *X* in the first column if you have never heard the word before.
Write an *X* in the second column if you have heard the word before.
In the last column, write what you think each word means.

Words to Know

Word	Have Not Heard Before	Have Heard Before	What I Think It Means
characteristic			
amphibian			
prey			

Write the answer.

What are the steps in the life cycle of a butterfly? ___________________________

____

____

____

Needs of Animals

Name ______________________________

Write an *X* below the living things.

1. _____

2. _____

3. _____

Write the answer.

4. Name the five needs of animals.

_______________ _______________ _______________

_______________ _______________

5. Animals live where their _______________ can be met.

6. Name two ways some animals change where they live to meet their needs.

by __________ __________ by __________ __________

7. How did God design animals and where they live to work together so the animals can survive and grow? Use the Bible's teaching from Genesis 1:28, 30 in your answer.

– Genesis 1:28, 30 –

God told Adam and Eve to care for the fish in the sea, the birds in the air, and all animals on land. He gave green plants as food for animals.

____

____

____

____

Classify Animals 1

Name ______________________

Complete the web about animals.

Animals

Animals With a ______________

Animals Without a ______________

Fish

- gills
- fins and ______________
- in ______________

Mammals

- lungs
- fur or ______________

Amphibians

- gills and ______________
- thin, smooth, wet ______________
- water and ______________

Birds

- lungs
- feathers
- 2 ______________
- most ______________

Reptiles

- lungs
- tough, dry, scaly ______________

Classify Animals 2

Name ______________________________

Complete the web about animals.

Animals

Animals With a ____________________

Fish

Birds

Mammals

Amphibians

Reptiles

Animals Without a ____________________

Insects

- 6 ______________
- 3 ______________ parts
- ______________ (usually)
- antennae

Spiders

- 8 ______________
- 2 ______________ parts
- ______________ wings
- ______________ antennae

Without Backbones

Name ________________________________

Compare and contrast insects and spiders by completing the chart.

Insects	Spiders
1. ______________ legs	2. ______________ legs
3. ______________ body parts	4. ______________ body parts
5. ______________ (usually)	6. no ______________
7. ______________	8. ______________ antennae

Write *W* for animals with a backbone. Write *WO* for animals without a backbone.

9. _____

10. _____

11. _____

12. _____

13. _____

14. _____

Fill in the blanks. One word will be used two times.

antennae	color	covering	eyes	pattern	webs

15. To help an insect know the world around it, an insect has two ________________.

16. To keep an insect safe, an insect has a hard ________________ outside its body.

17. To help keep insects safe, many insects have a ________________ or ________________ to help them hide.

18. To keep a spider safe, a spider has a hard ________________ outside its body.

19. To help spiders see, most spiders have eight ________________.

20. To help some spiders catch food, some spiders spin ________________.

Butterfly Life Cycle 1

Name ______________________

Follow the directions your teacher gives you.

Egg	Larva

Butterfly Life Cycle 2

Name ______________________________

Pupa	Adult

Butterfly Wings

Name ______________________________

Cut out the butterfly wings. Be careful to keep each tab connected to its wing.

Looking Ahead

Name ______________________

Write an *X* in the first column if you have never heard the word before.
Write an *X* in the second column if you have heard the word before.
In the last column, write what you think each word means.

Words to Know

Word	Have Not Heard Before	Have Heard Before	What I Think It Means
population			
community			
habitat			

Answer the question.

What living things can be found in a rainforest habitat? ______________________

Chapter 10

Bible Verse

Genesis 1:28

And God blessed them, and God said unto them, Be fruitful, and multiply, and replenish the earth, and subdue it: and have dominion over the fish of the sea, and over the fowl of the air, and over every living thing that moveth upon the earth.

Water Habitats

Name ____________________

Complete the organizer about water habitats.

Habitat	Fresh Water or Salt Water	Plants	Animals
Ocean			
Pond			
Wetland			

Study Guide

Name ______________________

Fill in the blanks.

community **habitat** **population**

1. All the living things of one kind that live in one area are a ______________.
2. All the living things in one area are a ______________.
3. A place where a living thing lives and finds food, water, and shelter is a

______________.

Use the picture to complete the sentence.

4. The turtle and fish are a community / population.
5. The fish are a community / population.

Mark the answer.

6. What does a dolphin depend on fish and squid for?

   ○ food ○ water ○ shelter

7. What are two ways tall grass meets the needs of ducks?

   ○ food ○ water ○ shelter

Match the plants and animals with their habitat.

_____ 8. cattails, frogs, turtles

_____ 9. dolphins, fish, seaweed

_____ 10. alligator, fish, grass

A ocean

B pond

C wetland

Write the answer.

11. God gave people the job of caring for plants and animals. Why is it important to learn and care about living things? Use the Bible truth from Genesis 1:28. ______________________

____

____

12. How does a habitat provide for the needs of plants and animals? ______________________

____

____

13. Would a fish be able to survive in a dry habitat? ______________________

____

____

14. How does a pond habitat meet the needs of a duck? ______________________

____

____

Land Habitats

Name ______________________

Complete the organizer about land habitats.

Habitat	Dry or Wet	Plants	Animals
Rainforest			
Woodland Forest	enough rain for plants to grow well		
Desert			
Savanna	dry and wet (rainy) seasons		
Tundra			

Home Sweet Home

Name ______________________________

Inquiry Skills
- Infer
- Communicate

Purpose

I want to model a habitat.

Materials

- ☐ 1 shoebox
- ☐ craft supplies

Procedure

☐ 1. Choose the habitat you want to model.

☐ 2. Record your habitat choice on the information card.

☐ 3. Use online resources or books to find out about the habitat.

☐ 4. Record the information about the habitat on the information card.

☐ 5. Plan the living and nonliving things to include in your habitat model. You must have at least two kinds of animals and one kind of plant.

☐ 6. Think about the materials you will use to represent the living and nonliving things. You may use real rocks and plants. You may draw pictures of animals and plants to use in your model.

☐ 7. Sketch your plans for the habitat. Draw each nonliving and living thing that you plan to use.

☐ 8. Label each item with what you plan to use to represent it.

☐ 9. Cut out your information card.

☐ 10. Attach the information card to the box.

☐ 11. Present your model and share what you have learned about the habitat.

Home Sweet Home

Name ______________________________

Use the Student Edition and other resources to complete the information on the card. Cut out your information card.

Information Card

____

This kind of habitat is found ______________________________.

Some nonliving things found here are ______________________________.

Two animals that live in this habitat are ____________________ and ______________________________.

One plant that lives in this habitat is ______________________________.

Something interesting about this habitat is that ______________________________
______________________________.

Home Sweet Home

Name ______________________________

Sketch the plans for your habitat model. Label each item with what you plan to use to represent it.

Habitat Model Sketch

Conclusions

1. How do your living things survive in their habitat?

____

____

____

2. Why could this habitat meet or not meet your needs if you lived there?

____

____

____

Looking Ahead

Name ______________________________

In the first column, write what you already know about fossils.
In the second column, write what you want to know about fossils.
The last column will be filled in after Lessons 79–80.

Fossils

What I ***Know*** about fossils	What I ***Want*** to know about fossils	What I have ***Learned*** about fossils

Answer the question.

If you have a biblical worldview, what do you believe about Creation? ______________________________

____

____

Chapter 11

Bible Verses

Genesis 1:24–27, 31

24 And God said, Let the earth bring forth the living
creature after his kind, cattle, and creeping thing,
and beast of the earth after his kind: and it was so.
25 And God made the beast of the earth after his
kind, and cattle after their kind, and every thing that
creepeth upon the earth after his kind: and God saw
that it was good.
26 And God said, Let us make man in our image, after
our likeness: and let them have dominion over the
fish of the sea, and over the fowl of the air, and over
the cattle, and over all the earth, and over every
creeping thing that creepeth upon the earth.
27 So God created man in his own image, in the
image of God created he him; male and female
created he them.
31 And God saw every thing that he had made, and,
behold, it was very good. And the evening and the
morning were the sixth day.

Genesis 6:19–20, 22

19 And of every living thing of all flesh, two of every
sort shalt thou bring into the ark, to keep them alive
with thee; they shall be male and female.
20 Of fowls after their kind, and of cattle after their
kind, of every creeping thing of the earth after his
kind, two of every sort shall come unto thee, to keep
them alive.
22 Thus did Noah; according to all that God
commanded him, so did he.

What's the Difference?

Name ________________________

Complete the chart by comparing and contrasting evolution and Creation.

by chance without God	during the flood	flood	God spoke
over many, many years	many times	perfect	slowly

Evolution	Creation
1. The earth formed ______________ ______________.	2. ______________ and created the earth.
3. The earth changed ______________. Death has always been a part of the world.	4. God made the earth ______________. The earth changed after Adam sinned and death became a part of the world.
5. The changes to the earth were made ______________ over a long, long time.	6. Man's sin caused God to send a ______________ that covered the whole earth.
7. Fossils formed slowly ______________ ______________.	8. Most fossils formed quickly ______________.

Read the story. Answer the questions.

Two scientists observe the stars. One scientist believes God created the stars. The other scientist believes the stars formed by chance without God.

9. Why do the two scientists infer something different about the stars that they are observing? ______________________________

____

____

____

____

10. Which scientist has a biblical worldview?

   ○ The scientist who believes God created the stars.

   ○ The scientist who believes the stars formed by chance.

EXPLORATION

Following Clues

Name ______________________________

Inquiry Skills

- Observe
- Measure
- Infer

Materials

- ☐ animal tracks
- ☐ ruler or meter stick
- ☐ colored pencils or crayons

Purpose

I want to study clues like a scientist.

Procedure

☐ 1. Look at the trails of animal tracks. Tracks are the footprints left behind by an animal.

☐ 2. Choose a trail of tracks to follow. Circle the animal tracks you will follow.

☐ 3. Imagine you are a scientist who discovered the trail. Look for clues about the animal's size, speed, and actions as you follow and study the animal's tracks.

☐ 4. Measure the length of one set of footprints using a ruler or meter stick.

Length of footprints: __________ cm

☐ 5. Measure the distance between two sets of animal footprints, or tracks, using a ruler or meter stick.

Distance between prints: __________ cm

☐ 6. Observe and infer about the animal that made the tracks.

- Does the animal have 2 legs or 4 legs? ______________ legs
- If the animal has 4 legs, can you tell the difference between the front feet and the back feet?

  ○ yes ○ no

- Is the animal big or small?

  ○ big ○ small

- How many toes do the front footprints show? ______________
- How many toes do the back footprints show? ______________
- Do the tracks show any webbing between the toes?

  ○ yes ○ no

- Try to move like the animal moved to make the trail of tracks.

  Do you think it walked, hopped, or ran? ______________

- In the space below, draw a picture of what you think the animal looks like and what it was doing when it left the tracks.

EXPLORATION

Following Clues

Name ______________________________

Conclusions

1. Find others who chose the same trail as you. Compare what was observed and inferred.
2. Did everyone in your group observe the same tracks?

   O yes O no

3. Did each of you infer the same things about the animal that left the tracks?

   O yes O no

4. How do you know that you inferred different things about the animal? ______________

   ____

5. Why do you think your drawings are different? ______________________

   ____

   ____

6. Can you know everything about an animal (what it looks like, where it lives, what it eats, where it was going, where it has been) by looking at its tracks?

   O yes O no

7. What information from science would you need in order to draw the right conclusions about the animal from the tracks it left? ______________________

   ____

   ____

Study Guide

Name ________________________________

Review

Complete the definitions. If needed, use the Glossary.

1. A worldview is a way a person thinks about the ________________.
2. **Creation** is the act by which ________________________ to make all things.
3. Evolution is the belief that the earth and space formed by ________________________.
4. A **paleontologist** is a scientist who studies ________________.
5. A **fossil** is the ________________ of a living thing that has died.

Write *C* if the statement is a belief of Creation.
Write *E* if the statement is a belief of evolution.

_____ 6. The earth changed many times.

_____ 7. God made the earth perfect.

_____ 8. The earth changed after Adam sinned.

_____ 9. Changes to the earth were made slowly over many, many years.

_____ 10. Man's sin caused God to send a flood that covered the whole earth.

_____ 11. Fossils formed over many, many years.

_____ 12. Most fossils formed quickly during the Flood.

Match the kind of fossil with the picture.

A

B

C

_____ 13. mold

_____ 14. cast

_____ 15. amber

Kinds of Fossils

amber   cast   frozen   mold   petrified   trace

Write *T* if the statement is true. If the statement is false, write the correct kind of fossil for the underlined words.

______________ 16. A <u>mold</u> fossil formed when mud hardened around a living thing that died and rotted, leaving its shape in the rock.

______________ 17. A <u>cast</u> fossil forms when a living thing leaves something behind, such as a footprint.

______________ 18. <u>Amber</u> is hardened tree sap.

______________ 19. A <u>petrified</u> fossil is a living thing that turned to rock.

______________ 20. A <u>trace</u> fossil formed in a place where the land is always frozen.

______________ 21. A <u>frozen</u> fossil is a copy of the living thing. It was made when soil filled the empty space of a mold and the soil turned to rock.

Read the story. Write the answer.

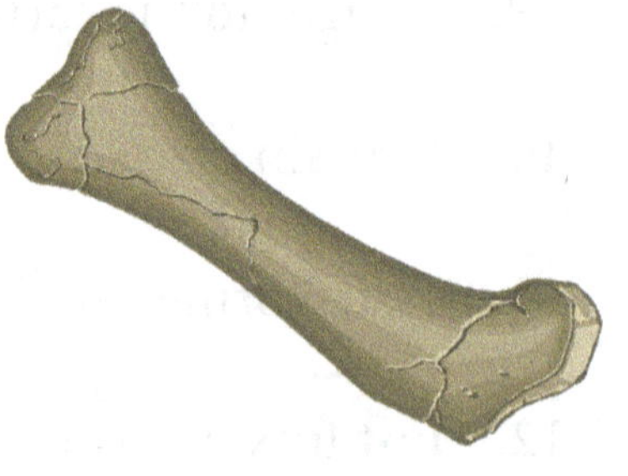

Two scientists observe a large dinosaur fossil. One scientist believes that the fossil formed quickly as a result of the Flood. The other scientist believes that the fossil formed over many, many years.

22. Why do the two scientists see the same fossil, but believe something different about how the fossil formed? ______________________________

____

____

____

____

Plant and Insect Fossils

Name ______________________________

Match the leaf fossil with the modern leaf.

_____ 1.

_____ 2.

_____ 3.

_____ 4.

maple A

willow B

oak C

palm D

Match the insect fossil with the modern insect.

_____ 5.

_____ 6.

_____ 7.

_____ 8.

dragonfly A

butterfly B

bee C

ant D

Write the answer.

9. What do plant and insect fossils tell us about life on the earth at the time of the Flood?

____

____

____

10. Why are some plants and insects found only as fossils? _______________________

____

Dinosaurs

Name ____________________

Sir Richard Owen was an English scientist who found some fossils. He wanted to describe what he found. He formed the word *dinosaur* and began using it in 1841. Then other scientists began using *dinosaur* to describe certain kinds of fossils.

The Bible was translated into English hundreds of years before Sir Owen lived. So the word *dinosaur* is not in our Bibles. However, there are some Bible verses that might be talking about this type of animal.

Read Job 40:15–24 to answer the questions.

1. What is this animal's name? (verse 15)

   ____

2. Who did God make at the same time that He made this animal? (verse 15)

   ____

3. What does this animal eat? (verse 15)

   ____

4. What is this animal's tail like? (verse 17)

   ____

5. What metals are this animal's bones like? (verse 18)

   ____

We do not know of an animal living now that fits this description. We can tell from the way Job describes this animal that it was very big and had a large tail. It might have been a dinosaur!

Study Guide

Name ______________________________

Review

Complete the definition. If needed, use the Glossary.

1. **Extinct** means that none of the plants or animals of that kind are still

   __________________.

If the statement is something that we can know by observing dinosaur fossils, circle the word *can*. If the statement is something scientists cannot know about dinosaurs by observing a fossil, circle the word *cannot*.

2. Dinosaur fossils can / cannot tell us that some dinosaurs laid eggs.

3. Dinosaur fossils can / cannot tell us the color of the dinosaur's skin.

4. Dinosaur fossils can / cannot tell us the size of a dinosaur.

5. Dinosaur fossils can / cannot tell us what kind of food the dinosaur ate.

6. Dinosaur fossils can / cannot tell us what kind of weather was best for the dinosaur to survive and grow.

Write *S* if the statement is true of the *Stegosaurus*. Write *T* if the statement is true of the *Tyrannosaurus rex*.

______ 7. is about the size of a bus

______ 8. had short front legs and stood on its back legs

______ 9. had two rows of bony plates along its back

Write the answer.

10. What do plant and insect fossils tell us about life on the earth at the time of the Flood?

____

____

____

____

11. What is one reason dinosaurs became extinct? _______________

____

Write About It

José and Tom are playing with toy dinosaurs. José tells Tom that people and dinosaurs lived on the earth at the same time. Tom does not believe this. Tom says that dinosaurs lived on the earth a long, long time before people. He says that the dinosaurs died many, many years before people lived on the earth. José wants Tom to know the truth about people, dinosaurs, and the Flood.

Help José explain to Tom what the Bible says in Genesis 1 about the creation of dinosaurs and people.

____

____

Explain to Tom what the Bible says in Genesis 6:19–22 about animals and the Flood. Include dinosaurs in your explanation.

____

____

EXPLORATION

Bag of Bones

Name ______________________________

Inquiry Skills

- Observe
- Infer
- Communicate

Materials

- ▢ bag of "bones"
- ▢ large sheet of construction paper
- ▢ crayon or marker
- ▢ glue

Purpose

I want to learn more facts about a dinosaur. I want to make a model of a dinosaur skeleton.

Procedure

- ▢ 1. Search "dinosaur skeleton pictures" on the internet to find pictures of dinosaur skeletons.
- ▢ 2. Choose one dinosaur. Write its name on the top of your construction paper.
- ▢ 3. Draw an outline of the dinosaur.
- ▢ 4. Spread out the bag of "bones" on your construction paper.
- ▢ 5. Arrange the "bones" to form the skeleton of your dinosaur.
- ▢ 6. Glue the "bones" in place.
- ▢ 7. Label the dinosaur bones that you can identify. For example, label the skull, backbone, leg bones, and ribs.
- ▢ 8. Search the internet using the name of the dinosaur whose skeleton you built.
- ▢ 9. Under Internet Research, list what scientists have observed about the dinosaur.
- ▢ 10. Under Internet Research, list what scientists have inferred about the dinosaur.
- ▢ 11. Write a paragraph describing your dinosaur.

Internet Research

____
(Name of Dinosaur)

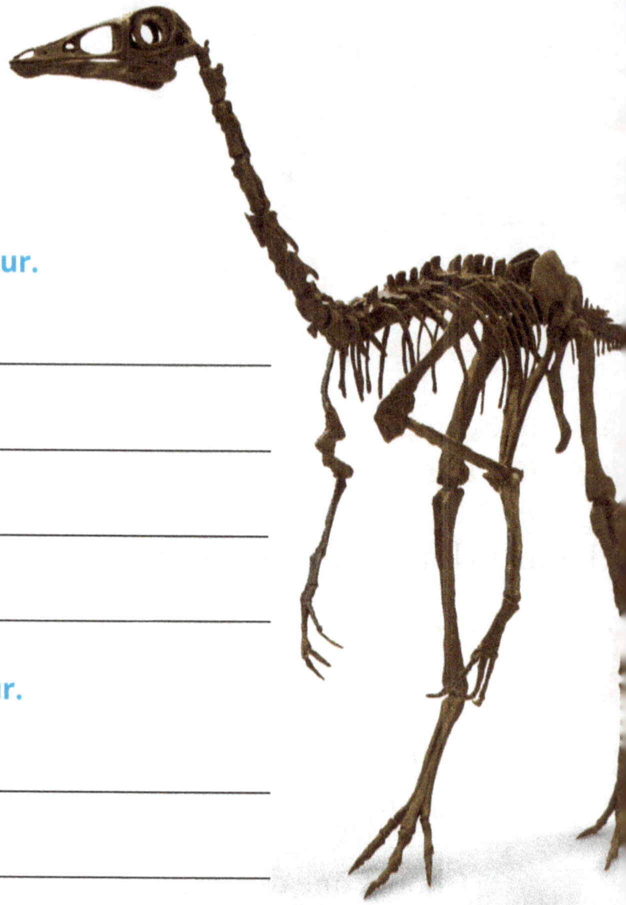

List what scientists have observed about this dinosaur.

____

____

____

____

List what scientists have inferred about this dinosaur.

____

____

____

____

I learned that my dinosaur ______________________________

____

____

____

____

____

____

____

EXPLORATION

Bag of Bones

Name ______________________________

Conclusions

1. According to the Bible, when did God create dinosaurs? ____________________

____

2. What event does the Bible describe where many dinosaurs died and were buried very quickly? ____

3. How do scientists infer what dinosaurs looked like? ______________________

____

4. Could you have made a different dinosaur using the same bag of bones?

   O yes   O no

5. What would you need to observe in order to know more about what your dinosaur looked like? ____

____

Looking Ahead

Name ______________________

Write an *X* in the first column if you have never heard the word before.
Write an *X* in the second column if you have heard the word before.
In the last column, write what you think each word means.

Words to Know

Word	Have Not Heard Before	Have Heard Before	What I Think It Means
system			
bone			
blood vessel			

Answer the question.

What is the job of the bones? ______________________

____

____

Chapter 12

Bible Verse

Psalm 139:14a

I will praise thee; for I am fearfully and wonderfully made:

Systems Organizer

Name ______________________

Write the name of one system in each box of the web.
Write the body parts that belong with each system on the web.

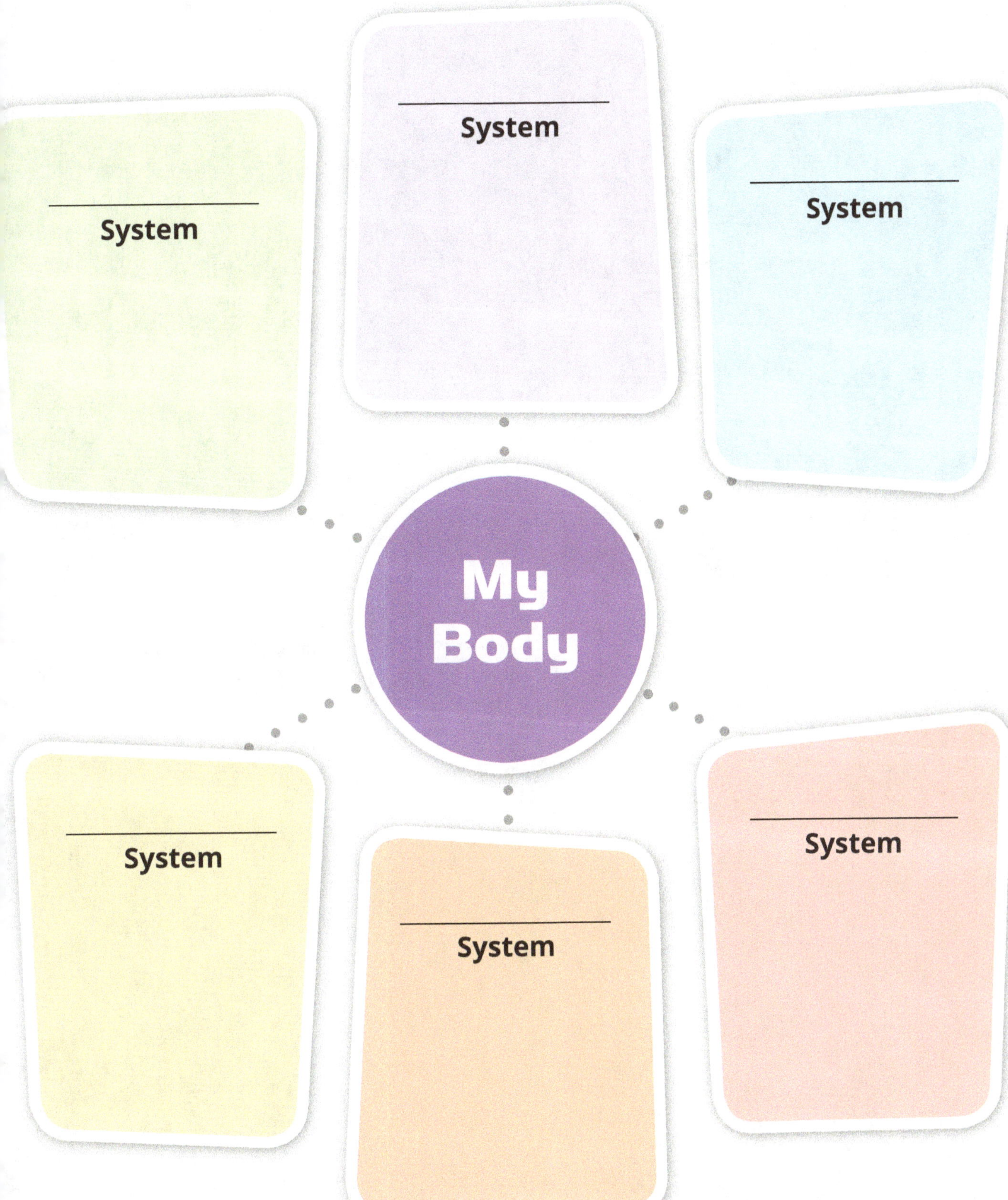

Skeletal and Muscular Systems

Name ______________________

Complete the definition. If needed, use the Glossary.

1. A **system** is a group of body parts that work ______________ to do a job.

2. A **bone** is a ______________ part of the body.

3. A **muscle** makes the ______________ and other body parts ______________.

Answer the question.

4. What are three jobs that bones have?

____

____

____

Mark the answer.

5. Which set of bones protects your heart and lungs?

○ ribs ○ skull ○ spine

6. Which bone protects your brain?

○ ribs ○ skull ○ spine

7. What muscle in your body works all the time, whether you think about it or not?

○ arm muscle ○ heart ○ leg muscle

Read the sentences. Write the answer.

People are the most important part of God's creation. We are more important than the sun, moon, and stars are to God. We are more important than plants and animals.

8. How do you know that people are the most important part of God's creation? Use the Bible's teaching in your answer. ______________________________

____

____

Study Guide

Name ______________________

Review

Complete the definition. If needed, use the Glossary.

1. A **system** is group of body parts that work ______________ to do a job.
2. **Bones** are the ______________ parts of the human body.
3. A **muscle** is the part of the human body that makes a bone or another body part ______________.
4. **Blood vessels** are tubes that carry ______________ to all parts of the body.
5. The **heart** is the powerful ______________ that pumps blood through the body.
6. The **lungs** are the body parts that fill with ______________ when you breathe in.

Write the body part in the system box that the part belongs to. You will use two words two times.

blood vessels	bones	diaphragm	heart	lungs	muscles

Skeletal System	Muscular System

Circulatory System	Respiratory System

Fill in the blanks.

in	out	pulls down	pushes up

15. The lungs fill with air when you breathe ______________.

16. Air leaves the lungs when you breathe ______________.

17. When the diaphragm __________ __________, the lungs fill with air.

18. When the diaphragm __________ __________, air leaves the lungs.

Mark all correct answers.

19. What two body parts work together to help the body move?

   ○ muscles   ○ blood vessels   ○ bones

20. Your heart is about the same size as your ___.

   ○ head   ○ foot   ○ fist

Answer the question.

21. How did God design the respiratory system and the circulatory system to work together?

____

____

Study Guide

Name ______________________________

Write About It

Sara and Emma are watching a TV program about wild animals. Sara tells Emma that animals and people are the same. Emma knows that animals and people are different.

What should Emma tell Sara to help her understand that people and animals are not the same?

____

____

____

____

____

Nervous and Digestive Systems

Name ______________________

Match the definition with the word.

A brain
B nerve
C spinal cord

_____ 1. controls everything in the body

_____ 2. connects the brain to the rest of the body

_____ 3. a tiny pathway that carries a message to the brain

Fill in the missing words to show the path that food travels through the digestive system.

large intestine **mouth** **small intestine** **stomach**

4. The place where food starts to digest is in the ______________.

5. The bag of muscle that mixes food with the digestive juices is the ______________.

6. The very long tube with thin walls where nutrients pass to the blood vessels is the ________ ________.

7. The last part of the digestive system where water is taken in is the ________ ________.

Label the parts of the nervous system.

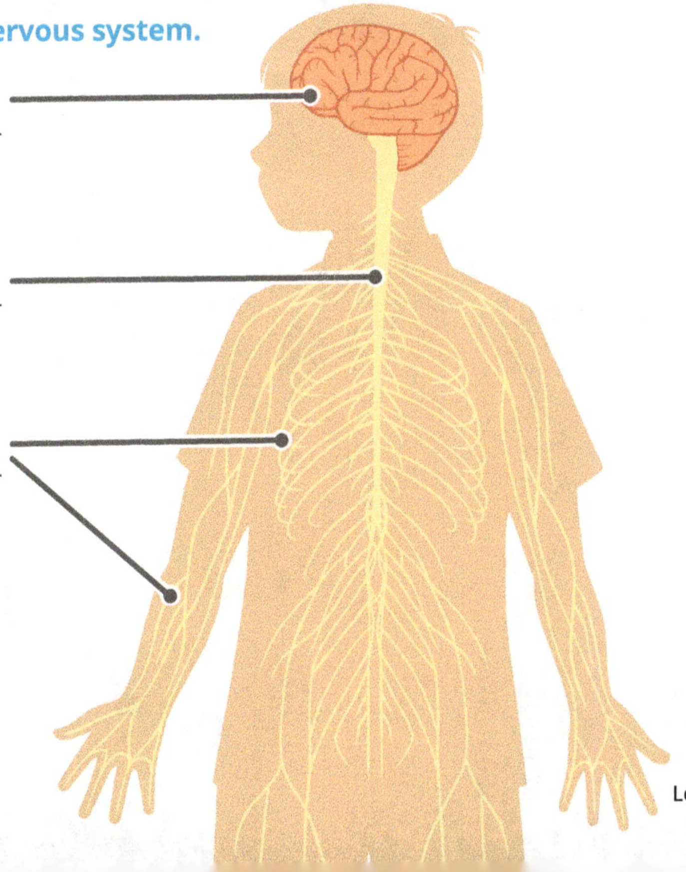

8. ______________

9. ______________

10. ______________

Label the parts of the digestive system.

11. ____________

12. ____________

13. ____________

14. ____________

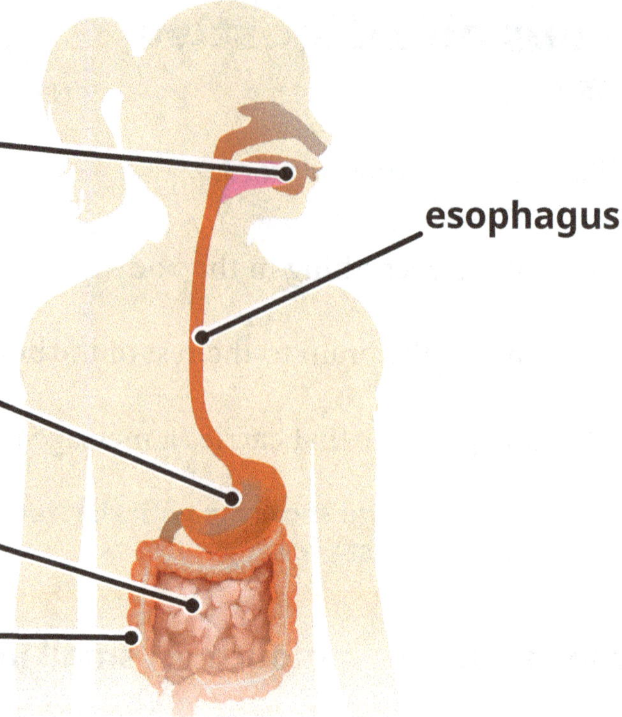

Answer the question.

15. How did God design the skeletal system to protect the nervous system? ____________

____

____

____

Being Healthy

Name ___________________________

Eating the right kinds of food is important. Food can be classified into different groups. Cut out the pictures on page 237. Glue them to the correct food group on the diagram.

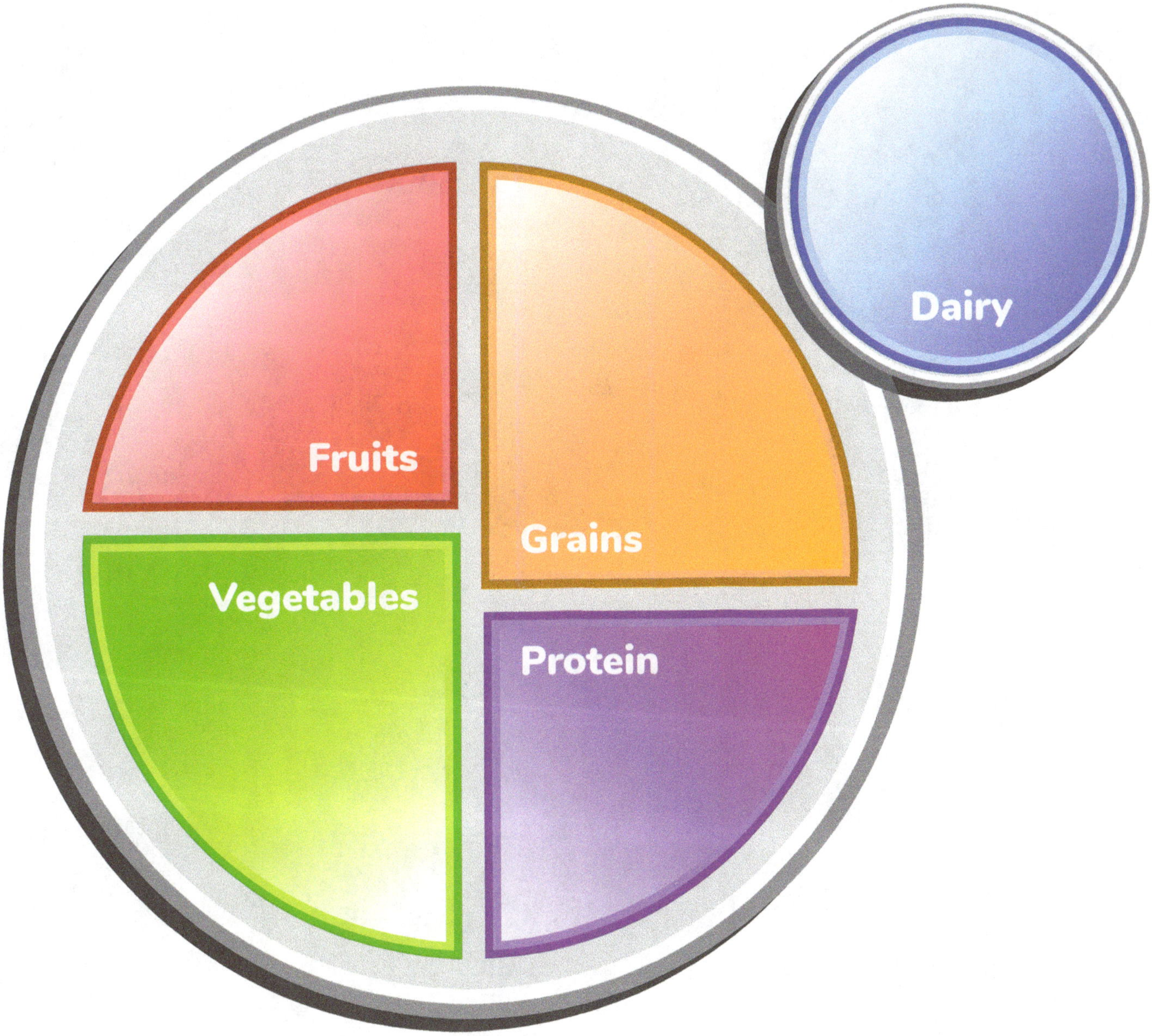

Healthy Foods

Name ______________________________

milk yogurt corn

ham cheese bread

broccoli turkey grapes

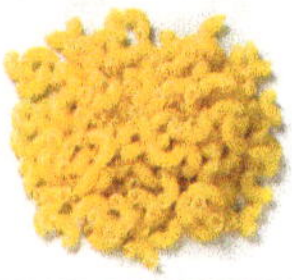  

pasta lettuce carrots

apple banana crackers

eggs nuts cereal

Healthy Eating Plan

Name ______________________________

Plan one day of healthy eating. Use the "Being Healthy" diagram to help you plan three meals and a snack. Write your food choices on the chart.

One Day Healthy Eating Plan

	Breakfast	Lunch	Snack	Dinner
Fruit				
Grains				
Protein				
Vegetables				
Dairy				

Healthy Moving

Circle three pictures that show ways for the body to get exercise.

Study Guide

Name ______________________________

Write the body part in the system box that the part belongs to.

brain	large intestine	mouth	nerves
small intestine	spinal cord	stomach	

Nervous System	Digestive System

Number the body parts to show the path of digestion in order.

8. ______ stomach

______ large intestine

______ mouth

______ small intestine

Write *T* if the statement is true. If the statement is false, write the correction for the underlined words.

______________ 9. The <u>brain</u> controls everything in the body.

______________ 10. A tiny pathway that carries a message to the brain is the <u>spinal cord</u>.

______________ 11. The <u>stomach</u> is a bag of muscle that mixes food with digestive juices.

Place a ✓ on the line if the food is a healthy eating choice.

_____ 12. soda

_____ 13. hard-boiled egg

_____ 14. carrot sticks

_____ 15. turkey sandwich

_____ 16. potato chips

_____ 17. yogurt with fruit

Match the food with the food group.

_____ 18. cheese

_____ 19. grapes

_____ 20. broccoli

_____ 21. ham

_____ 22. cereal

A grains

B protein

C vegetables

D fruits

E dairy

Circle the picture that shows a good way to get exercise.

23.

Study Guide

Name ____________________

Write the answer.

24. How has the skeletal system been designed by God to protect the nervous system?

____

____

25. Why is it important to eat healthy and exercise every day? ____________________

____

____

Write a prayer of praise to thank God for His design of the human body systems.

26. ____________________

____

____

____

____

____

____

____

____

EXPLORATION

Mapping My Body

Name ______________________

Inquiry Skills
- Infer
- Communicate

Materials

- ☐ large sheet of banner paper
- ☐ 1 sheet of red construction paper
- ☐ scissors
- ☐ glue
- ☐ brain shape from page 247
- ☐ lungs shape from page 249
- ☐ stomach shape from page 251
- ☐ crayons or markers

Purpose

I want to make a map of some of my body systems.

Procedure

- ☐ 1. Lie flat on a large sheet of paper. Ask your partner to trace the outline of your body.
- ☐ 2. Use the diagrams of the body systems from this chapter to help you place and label the parts.
- ☐ 3. Cut out the **brain** shape and glue it to your map.
- ☐ 4. Label the brain.
- ☐ 5. Draw the **spine** on your map.
- ☐ 6. Label the spine.
- ☐ 7. Draw a **heart** on the red construction paper. Remember to make it about the size of your fist. Cut it out and glue it to your map.
- ☐ 8. Label the heart.
- ☐ 9. Cut out the **lungs** shape and glue it to your map.
- ☐ 10. Label the lungs.
- ☐ 11. Cut out the **stomach** shape and glue it to your map.
- ☐ 12. Label the stomach.

- [ ] 13. Draw a face on your map with eyes, a nose, and a **mouth**.
- [ ] 14. Label the mouth.
- [ ] 15. Draw a tube from the mouth to the stomach.
- [ ] 16. Use your map to tell others about the body systems. Share what you can do to keep your body healthy.

Conclusions

1. What systems are shown on your body map? ____________________

____

2. What are some ways you can keep your body healthy? ____________________

____

3. How does your body map help you to understand Psalm 139:14?

____

____

____

____

____

— Psalm 139:14a —

I will praise thee; for I am fearfully and wonderfully made.

EXPLORATION

Mapping My Body

Name ___________________________

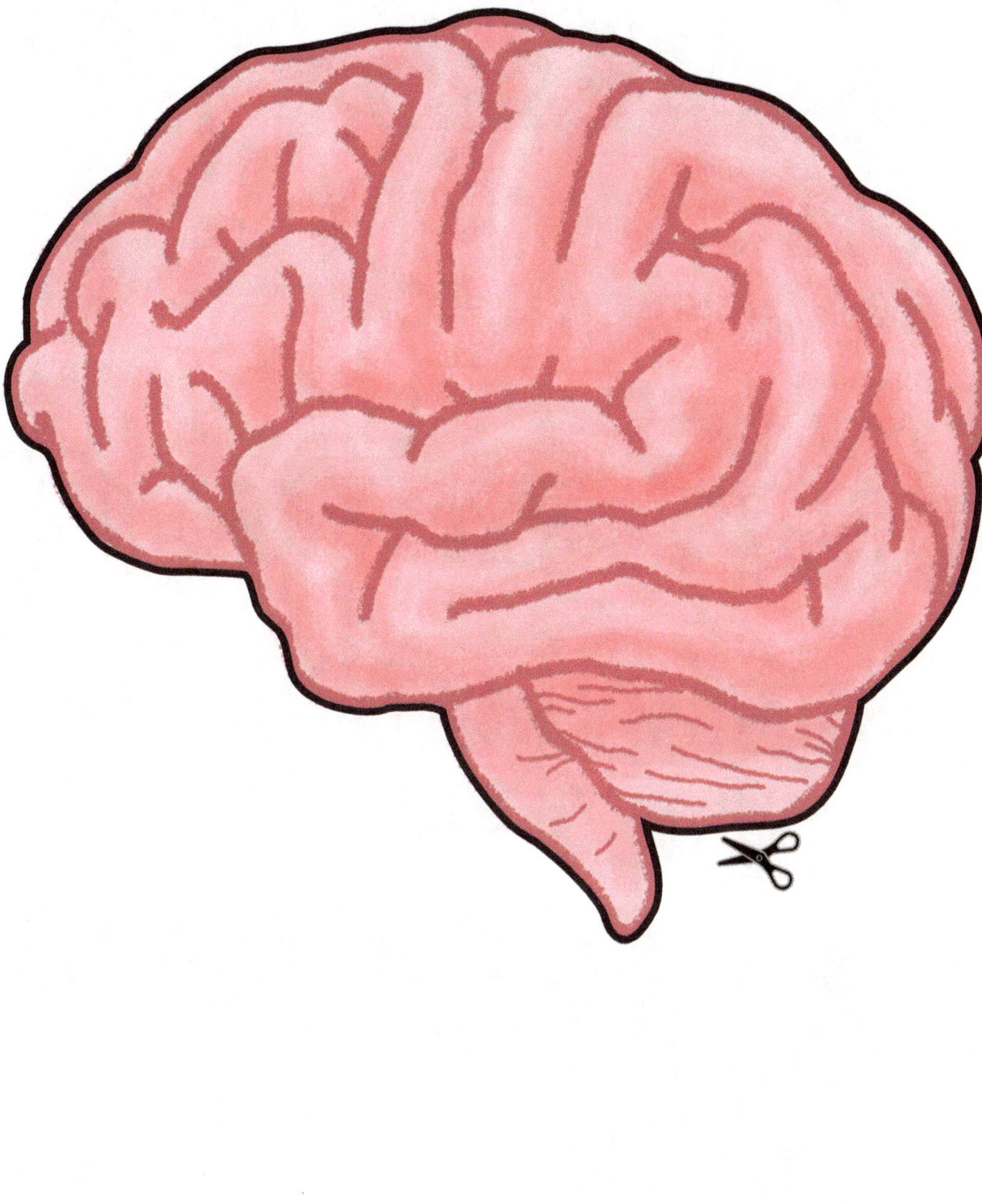

EXPLORATION

Mapping My Body

Name ______________________________

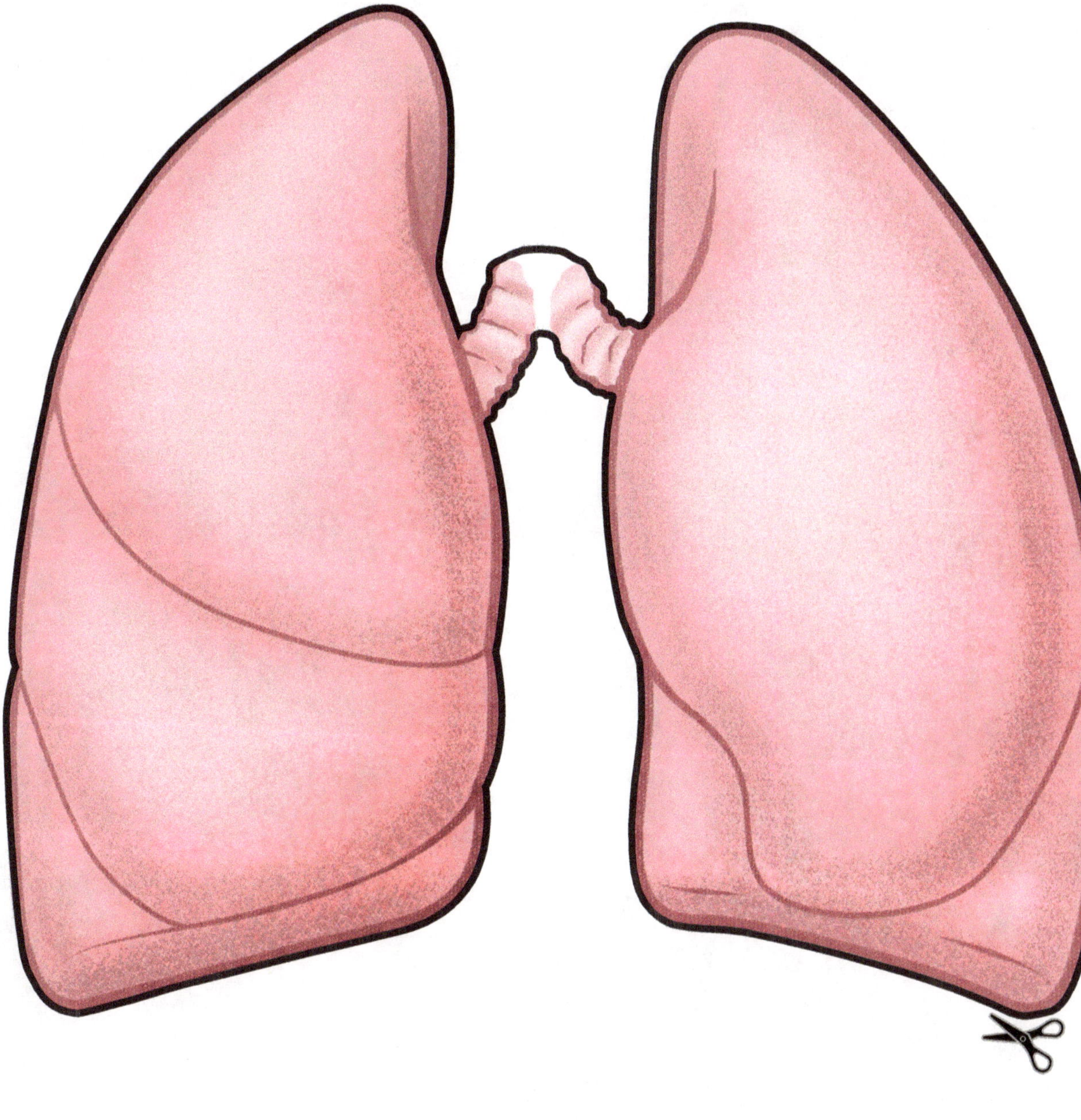

EXPLORATION

Mapping My Body

Name ______________________________

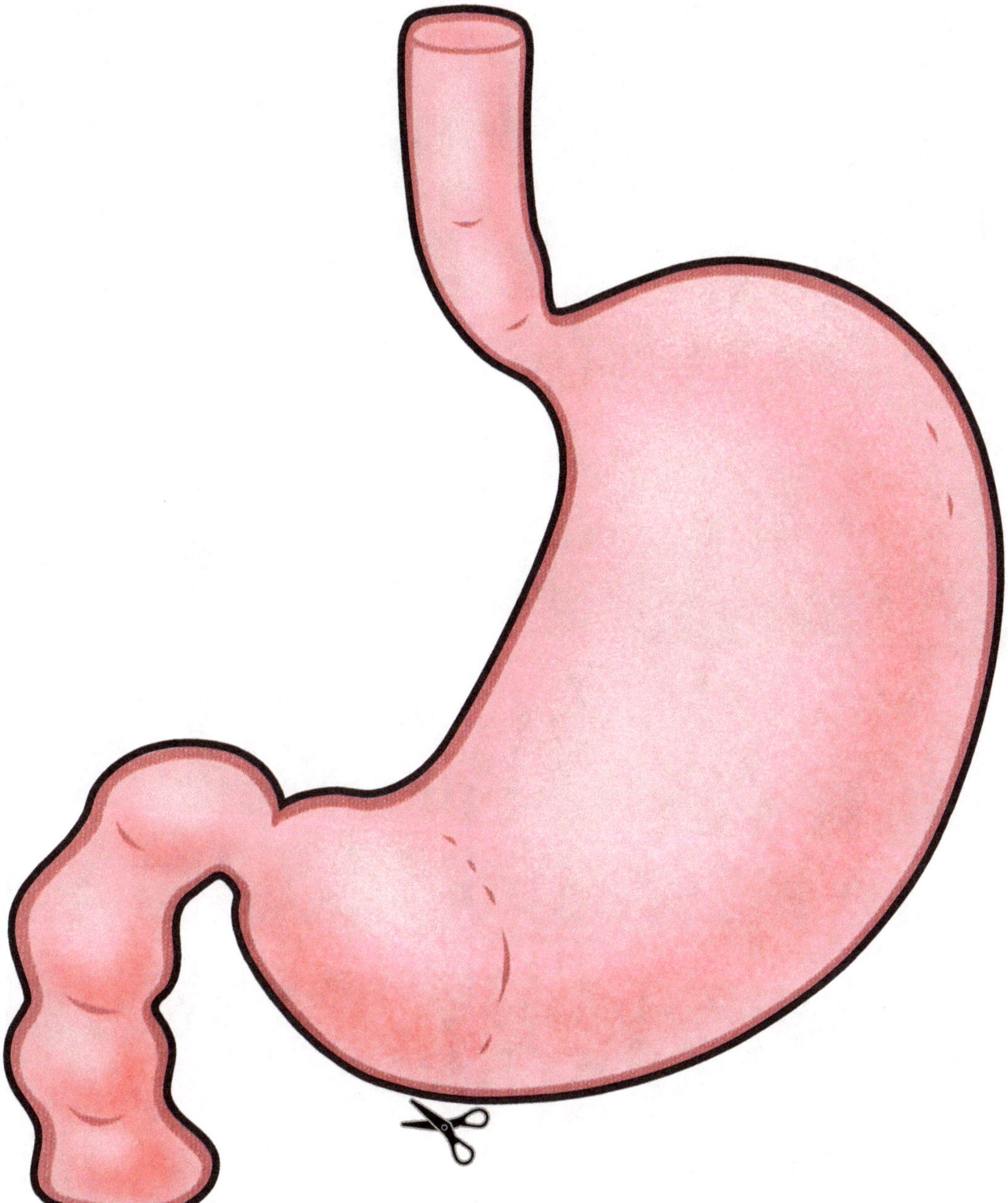

Photo Credits

Key (t) top; (c) center; (b) bottom; (l) left; (r) right;

Front (butterfly) suns07butterfly/Shutterstock.com; (leaf insect) © Isselee | Dreamstime.com; (Tyrannosaurus Rex) DisneyTD/TurboSquid; **back** Subsociety/iStock/Getty Images Plus/Getty Images

Chapter 1
1tl, **7**br © iStock.com/triocean; **1**tr © iStock.com/baona; **1**bl cmannphoto/iStock/Getty Images Plus/Getty Images; **1**br © iStock.com/mdesigner125; **2** "Spinophorosaurus holotype" by Remes K, Ortega F, Fierro I, Joger U, Kosma R, et al./Wikimedia Commons/CCA By-SA 2.5 Generic; **7**tl © iStock.com/Photo-Dave; **7**tcl © iStock.com /Zadas_Photography; **7**tcr Asia Images Group /Shutterstock.com; **7**tr Noel Hendrickson/Photodisc/Getty Images; **7**bl tomwang/123RF; **7**bc pornpan chaiu-dom /Shutterstock.com; **8**t © iStock.com/AfricaImages; **8**ct © piai/Fotolia.com; **8**c BJU Photo Services; **8**cb Steve Debenport/E+/Getty Images; **8**b deyangeorgiev/iStock /Getty Images Plus/Getty Images; **14** Sergey Novikov /Shutterstock.com

Chapter 2
23tl niwat chaiyawoot/Shutterstock.com; **23**tcl Gchapel /Shutterstock.com; **23**tcr Yellow Cat/Shutterstock.com; **23**tr Africa Studio/Shutterstock.com; **23**bl sbayram/E+ /Getty Images; **23**bcl topseller/Shutterstock.com; **23**bcr, **25**tr irin-k/Shutterstock.com; **23**br Eric Isselee /Shutterstock.com; **25**tl wabeno/iStock/Getty Images Plus /Getty Images; **25**tcl chengyu/Shutterstock.com; **25**tcr showcake/Shutterstock.com; **25**bl Yuriy Seleznev /Shutterstock.com; **25**bcl Stepan Bormotov/Shutterstock .com; **25**bcr inerika/Shutterstock.com; **25**br Feng Yu /Shutterstock.com; **35**l VladaKela/Shutterstock.com; **35**c Anton Starikov/Shutterstock.com; **35**r pukach /Shutterstock.com; **39** BJUPress

Chapter 3
46, 53 BJUPress; **58**t Nataliia K/Shutterstock.com; **58**c BrianWancho/Shutterstock.com; **58**b Floortje/iStock /Getty Images Plus/Getty Images

Chapter 4
63tl joebelanger/iStock/Getty Images Plus/Getty Images; **63**tr ElementalImaging/E+/Getty Images; **63**cl MCCAIG /iStock/Getty Images Plus/Getty Images; **63**cr Graham Oliver/MediaBakery; **63**bl Peerayot/Shutterstock.com; **63**br Erickson Stock/Shutterstock.com; **69** BJU Press; **75**tl, **75**tr, **75**bl © 2009 JupiterImages Corporation; **75**br Ben Girardi/Getty Images; **76**tl David Papazian/Shutterstock .com; **76**tr Pawan S/Shutterstock.com; **76**bl Slip-X Solutions; **76**br vvoe/Shutterstock.com; **79** DoD/SSGT JEFFREY A. WOLFE; USAF

Chapter 5
96 © Ivan Kmit | Dreamstime.com

Chapter 6
103 Cavan Images/Cavan/Getty Images; **104, 108** Cartesia Software; **109** David Nunuk/All Canada Photos/Getty Images; **111** Wanannc/Shutterstock.com; **113** AY Amazefoto/Shutterstock.com; **118** Martin Rietze /Media Bakery; **119** Dean Fikar/Moment/Getty Images

Chapter 7
135t © iStockphoto.com/nojustice; **135**c Chanwoot_Boonsuya/Shutterstock.com; **135**b Sean D /Shutterstock.com

Chapter 8
143l Nejron Photo/Shutterstock.com; **143**cl Issaurinko /iStock/Getty Images Plus/Getty Images; **143**cr Ruslan Ivantsov/Shutterstock.com; **143**r Martin Mistretta/The Image Bank/Getty Images Plus/Getty Images; **144** tom viggars/Alamy Stock Photo; **150**l Vasin Lee/Shutterstock .com; **150**c Jackan/Shutterstock.com; **150**r Andy Roberts /OJO Images/Getty Images; **151** BJU Press

Chapter 9
156l Jurgen & Christine Sohns/Getty Images; **156**c sunara /E+/Getty Images; **156**r Nigel Jarvis/Shutterstock.com; **162**tl © iStockphoto.com/oneclearvision; **162**tr © iStockphoto.com/Vitalii Hulai; **162**bl © iStockphoto .com/Steve Snyder; **162**br © iStockphoto.com/forestpath

Chapter 10
185t Daniela Dirscherl/WaterFrame/Getty Images Plus /Getty Images; **185**bl Danita Delimont/Gallo Images /Getty Images; **185**br Arto Hakola/Shutterstock.com; **192** Ronnie Howard/Shutterstock.com; **193** abd.halim hadi/ Moment/Getty Images

Chapter 11
206 John Davis/Stocktrek Images/Getty Images; **211**l Beker/Shutterstock.com; **211**c Breck P. Kent /Shutterstock.com; **211**r Gabbro/Alamy Stock Photo; **213**lt John Cancalosi/Photolibrary/Getty Images Plus/Getty Images; **213**lct Gilles Mermet/Science Source; **213**lcb Wild Horizon/Universal Images Group/Getty Images; **213**lb BIOPHOTO ASSOCIATES/Science Source/Getty Images Plus/Getty Images; **213**rt Ken Hurst/Alamy Stock Photo;

213rct Maryna Kulchytska/Shutterstock.com; **213**rcb cws_design/iStock/Getty Images Plus/Getty Images; **213**rb Dzmitrock87/iStock/Getty Images Plus/Getty Images; **214**lt The Natural History Museum, London/Science Source; **214**lct John Cancalosi/Photolibrary/Getty Images Plus/Getty Images; **214**lcb Custom Life Science Images/Alamy Stock Photo; **214**lb Jean E. Roche/NPL/Minden Pictures; **214**rt Huntstock, Inc/Getty Images; **214**rct Norbert Liesz/500px /500Px Plus/Getty Images; **214**rcb Fauzan Maududdin /EyeEm/Getty Images; **214**rb twomeows/Moment/Getty Images; **218** Hoberman Collection/Universal Images Group /Getty Images; **220** breckeni/E+/Getty Images

Chapter 12

230 Henglein and Steets/Cultura/Getty Images; **235** LPETTET/DigitalVision Vectors/Getty Images; **237** (milk) Anton Starikov/Shutterstock.com; **237** (yogurt) Pixel-Shot /Shutterstock.com; **237** (corn) chengyuzheng/iStock/Getty Images Plus/Getty Images; **237** (ham) Tetra Images/Getty Images; **237** (cheese) Marat Musabirov/iStock/Getty Images Plus/Getty Images; **237** (bread) LauriPatterson/iStock/Getty Images Plus/Getty Images; **237** (broccoli) pidjoe/iStock /Getty Images Plus/Getty Images; **237** (turkey) Africa Studio /Shutterstock.com; **237** (grapes) Natthakan Jommanee /EyeEm/Getty images; **237** (pastas) Fotosoroka/Shutterstock .com; **237** (lettuce) mikroman6/Moment/Getty Images; **237** (carrots) Nattika/Shutterstock.com; **237** (apples) Zoonar RF/Zoonar/Getty Images Plus/Getty Images; **237** (banana) harmpeti/iStock/Getty Images Plus/Getty Images; **237** (crackers) bergamont/iStock/Getty Images Plus/Getty Images; **237** (eggs) OLEKSANDR PEREPELYTSIA/iStock /Getty Images Plus/Getty Images; **237** (nuts) Geo-grafika /Shutterstock.com; **237** (cereal) george tsartsianidis/iStock /Getty Images Plus/Getty Images